MES

A Technician's Real-World Guide to Manufacturing Execution Systems

MES

Kimberly Cousins

MES

A Technician's Real-World Guide to Manufacturing Execution Systems

First Edition

ISBN: 978-0-9816842-9-1

Cover design by Kimberly Cousins
Interior layout by Kimberly Cousins

Printed in the United States of America

Table of Contents

Introduction: Why You Need to Read This Book First 8

Part 1 9

What MES Really Is 9

Chapter 1 11

Where MES Sits 11

(The Part Rarely Explained) 11

Part 2 17

Understanding the Bigger Picture 17

Chapter 2 19

Unraveling the Layers of MES 19

Chapter 3 32

How ERP and MES Work Together 32

Part 3 45

Creating Processes and Production 45

Chapter 4 47

Building the Process: Operations, Recipes, and Tools 47

Chapter 5 63

What "Configuration" Really Means in MES 63

Part 4 75

A Day in the Life of an 75

MES Technician 75

Chapter 6 77

Office vs Shop Floolr Reality 77

Chapter 7 87

How MES Failures Actually Happen 87

Chapter 8 93

On-Call Shift Story: When Everything Breaks at Once.................. 93

Chapter 9 113

MES Technician vs MES Engineer: What's Actually Different? . 113

Chapter 10 131

Three Common MES Problems Every Technician Will Face....... 131

MES Quick Reference:... 137

Common Problems Checklist 137

Introduction: Why You Need to Read This Book First

Most books about Manufacturing Execution Systems talk about strategy, integration, and business value. That's all useful, if you're sitting in an office. This book explores a slightly different angle. This is about what it's like to actually use MES on the Shop Floolr—when the line is running, the clock is ticking, and the system doesn't always behave the way it should. It explains MES not as theory, but as real life viewpoint of the MES technician.

- What goes wrong
- How to fix it
- Why it matters to production

These are practical items to look out for and how to avoid the trap of becoming bogged down over processes and instead more in tune with actual production needs and how to forecast and resolve certain problems before the bottleneck effect becomes reality. Especially for those who are just becoming acquainted with MES, it's pertinent to learn the skills ahead of time so that when problems arise, you are aware of what caused it and how to swiftly resolve the issue or elevate to the next tier to locate a solution. Production often moves swiftly. Therefore, results matter. But this is something that is often not taught until faced with an unexpected technical issue for which no one gave any forethought.

Part 1
What MES Really Is

Chapter 1
Where MES Sits (The Part Rarely Explained)

People often confuse MES with a simple computer-based data entry tool, but it is much more than that. MES is often what separates volatile production environments from scalable, stable growth. When I first became acquainted with MES, I was working at a semiconductor manufacturing company where paper travelers—step-by-step instruction sheets—were used to create, document, and essentially drive the execution of processes on the floor.

Later, as the company moved toward automation, MES became an essential part of the production process. It was no longer optional. Staff had to quickly shift their mindset away from manual, paper-based step-by-step tracking and adapt to a more structured, system-driven approach designed to improve consistency and yield.

While some resisted the change, others embraced this more streamlined, data-driven approach to production. It reduced the likelihood of human error, but one tradeoff—what some considered a downside—was that it also reduced the ability to manually override metrics or decisions that didn't appear to make sense at first glance.

MES isn't something you fully understand from a manual. You learn it when production is backed up, a *lot* won't move, and three different people are asking you what's wrong—at the same time. One person thinks it's equipment. Another thinks it's the operator. And someone from management is already asking how long before things are back online. That's where MES lives—not in theory, but in the middle of pressure.

So... What is MES? (Without the Corporate Talk)

So, what is MES? At its core, a Manufacturing Execution System (MES) is the system that tracks, controls, and documents production on the factory floor in real time. That's the formal definition. Here's the real one: MES is the system that tells the factory what's happening, what should happen next, and what already went wrong.

It connects:

- Machines
- Operators
- Processes
- Data

And it turns all of that into something the business can track, measure, and act on.

Logging into the System (What It Really Means in MES)

Logging into MES is usually the first step of the day—but it's more than just entering a username and password. It's the point where the system decides: "Who are you, and what are you allowed to do?" In MES, access is not universal. It is structured, controlled, and intentionally limited. Depending on your level of access as a technician, you may be required to set user limitations of other users as well.

Accessing MES Based on Role and Permissions

When you log in, MES doesn't just open the system—it loads a **role-based environment**.

That means your access depends on your job function. For example:

- A technician may see operational screens, lots, and execution tools
- An engineer may see configuration, recipes, and system setup options
- A supervisor may see dashboards, production status, and reporting views
- An operator may only see execution steps tied to their station

You're not just logging into MES—you're accessing a version of the system tailored to your specific responsibilities. In a previous role as the sole MES technician, I had full access across all system levels. I could start lots, manage configurations, maintain recipes and system settings, access dashboards, pull KPI data from reports, and even monitor and control execution steps. I was also responsible for resetting passwords for users with expired or forgotten credentials. Depending on the company's needs, user roles were often assigned on an ad hoc basis.

Why MES Uses Role-Based Access

This design is intentional. MES systems control real production environments, so they must:

- Limit critical changes to authorized users
- Prevent accidental modifications to configuration

- Ensure data integrity across all users
- Separate execution from system design

In other words: Not everyone should be able to do everything. This protects both the system and the production process.

Why Different Users See Different Views

One of the first confusing things in MES is realizing: "Why does someone else see something I don't see?" As mentioned previously, different roles are shown different **views of the same system**.

Example:

A technician may see:

- Lot tracking screens
- Operation status
- Tool assignments
- Error messages and holds

An engineer may see:

- Recipe configuration
- Operation setup
- Tool definitions
- System logic and rules

A manager may see:

- KPIs

- Downtime reports
- Production efficiency dashboards

Same system. Different lens.

What This Means for You as a Technician

Understanding this early helps you avoid confusion later. Because when you see:

- A missing screen
- A locked field
- Or unavailable function

It's not necessarily a system issue. It may simply be: outside your role permissions.

Common Misunderstanding

New technicians often assume: "If I can't see it, something is wrong." But in MES, visibility is intentional. Not everything is meant to be visible to every user.

Key Insight

Logging in is not just system entry—it's identity assignment. MES immediately determines:

- What you can access
- What you can change
- What you can only observe

Closing Thought

In MES, access defines responsibility. Responsibility defines visibility. And visibility defines what part of the system you ar.e trusted to work in

Part 2
Understanding the Bigger Picture

Chapter 2
Unraveling the Layers of MES

Think of MES as the middle layer between two worlds: Enterprise Resource Planning (ERP), Manufacturing Execution System (MES), Factory Floor (Fab).

Top Layer (Business Systems i.e. ERP):

- Planning
- Orders
- Inventory
- Customer demand

Bottom Layer (Factory Floor):

- Equipment
- Tools
- Operators
- Physical production

MES sits in between and translates everything.

- The business says: *"Build 500 units."*
- MES says: *"Here's how, where, and in what order."*
- The floor executes it.
- MES tracks every step.

If MES is working well, everything flows. If MES is down or misbehaving, everything slows—or stops.

What MES Actually Does (From a Technician's View)

From your perspective as a technician, MES isn't some abstract system. It's the thing you interact with all day:

- Moving lots
- Checking statuses
- Resolving holds
- Fixing errors
- Verifying routes
- Making sure data is correct

You're not just "using" MES. You're maintaining the flow of production through it. And that's a big difference.

The Truth About MES Jobs

Here's something most job descriptions won't tell you:

MES is part technical skill, part problem-solving, and part pressure management. Because when something goes wrong:

- Production is waiting
- Operators are stuck
- Engineers need answers
- Managers want timelines

And you're in the middle of it. You are not always presented with clean problems to solve. Sometime you get:

- Incomplete information
- Conflicting explanations
- Systems that *should* work—but don't

Common Misconceptions About MES

Let's unravel a few misconceptions early:

➢ **"MES is just IT work"**

No.

You're not just fixing systems—you're supporting production. If you don't understand the manufacturing process, you'll struggle.

➢ **"The system tells you everything"**

Also *no*.

MES gives you data, but it doesn't give you judgment. You still have to interpret what you're seeing.

➢ **"If something breaks, it's always MES"**

No. This happens all the time. Something breaks in the line and many will assume that it is an MES issue.

Sometimes it's:

- Equipment

- Operator error
- Process issues
- Bad data upstream

Blaming MES is easy. Proving what's actually wrong is the real skill.

What No One Tells You

Here's the part most don't consider:

- You will feel lost in the beginning
- You won't understand every error message
- You'll fix things without fully knowing why they broke
- You'll learn patterns before you learn theory

And that's normal. MES isn't something you master in training. You build it through repetition, mistakes, and exposure.

Why This Role Matters More Than You Think

If MES fails, the impact isn't just technical—it's operational and financial:

- **Production slows**
  When MES goes down or lags, operators can't move lots, track progress, or complete transactions. Equipment may sit idle while teams wait for system access. In high-volume environments, even a few minutes of downtime can ripple into hours of lost output.

- **Data becomes unreliable**
  MES is the system of record. If data is missing, duplicated, or incorrect, it compromises traceability, quality records, and compliance. This isn't just messy—it can lead to failed audits, product holds, or even recalls.

- **Decisions get delayed**
  Engineers, supervisors, and leadership rely on MES dashboards and reports to make real-time decisions. If the data is late or wrong, decisions are either delayed or made based on bad information—both of which increase risk.

- **Money is lost**
  This is where everything connects. Lost production time, scrapped material, rework, delayed shipments, and poor decisions all translate directly into financial loss. In some facilities, a single hour of downtime can cost thousands—or more.

That's why your role matters. You're not just "support." You're the bridge between the system and the production floor. When MES runs smoothly, production flows, data is trusted, and the business performs. When it doesn't, everything feels it.

How and Why Money Is Lost When MES Fails

So where does that financial loss actually come from? Money isn't lost in one obvious way—it leaks out through multiple channels, often at the same time:

- **Idle equipment and labor (lost production time)**
  When MES is down or slow, operators can't move lots, log steps, or proceed with work. Machines sit idle, but they're still costing money—utilities, depreciation, and labor are

all still running. You're paying for production that isn't happening.

- **Scrap and rework (material loss)**
  If recipes, parameters, or process steps are incorrect or unavailable, products may be processed the wrong way. That can lead to scrap (total loss) or rework (extra time and cost to fix it). Either way, material and labor costs increase.

- **Delayed shipments (missed revenue or penalties)**
  When production slows, orders don't go out on time. This can result in missed delivery commitments, financial penalties, or even lost customers. In some industries, delays directly impact contractual revenue.

- **Bad data = bad decisions (hidden financial impact)**
  MES drives reporting. If the data is wrong or delayed, leadership may make decisions based on inaccurate information—like overproducing, underproducing, or misallocating resources. These mistakes quietly cost the company money over time.

- **Compliance and audit risks**
  In regulated industries, missing or incorrect MES data can fail audits. That can lead to fines, shutdowns, or product recalls—all of which are extremely expensive and damaging.

- **Firefighting instead of efficiency**
  When MES issues occur, teams shift from planned production to reactive problem-solving. Engineers, supervisors, and technicians spend time troubleshooting

instead of improving processes. That lost efficiency adds up quickly.

The Key Insight for Technicians

Money is lost any time:

- Time is wasted
- Materials are wasted
- People are waiting
- Or decisions are made with bad information

MES sits in the middle of all four.

That's why even small issues—like a stuck lot, a misconfigured recipe, or delayed data—can have a much bigger financial impact than they appear on the surface.

Real-Life Scenario: The Layer I Ignored (And Paid For Later)

When I first got into MES, I did what most technicians do. I stayed focused on what was directly in front of me. I was focused on processes, operations, recipes, and work instructions. I didn't take the necessary time to learn much about the ERP layer. I wish I would have considered that earlier when I first became aware of MES instead of having to learn from the school of hard knocks. After all, inventory and work orders must come from somewhere. It's good to understand the planning phase and the essential function of how MES and ERP cooperate:

- Processes
- Operations
- Recipes
- Work instructions

Everything that lived inside the MES environment felt like *the job*, so that's where I put my attention. And for a while, that worked. Until it didn't.

The Moment Things Stopped Making Sense

I started running into issues that didn't line up.

- Lots that wouldn't move—but no clear MES error
- Work orders that looked valid—but weren't processing correctly
- Inventory mismatches that MES couldn't explain
- Data that looked wrong, but not *broken enough* to trigger alarms

At first, I treated these like typical MES problems: Check the route. Check the operation. Check the system status. Everything looked fine. But production was still stuck.

What I Didn't Understand Yet

What I hadn't taken the time to learn was the layer *above* MES—the ERP system. When that gap showed up in real time, it was an immediate on-the-job training experience that heightened my understanding of cross functionality. Because here's the truth: MES doesn't create demand. It executes it. MES doesn't generate inventory. It consumes and tracks it. That all starts somewhere else.

The Missing Link: ERP

The ERP system is where:

- Work orders are created
- Inventory is defined and allocated
- Production demand is planned
- Materials are structured (BOMs)

By the time something reaches MES, it's already been decided, structured, and released upstream. So when something is wrong in ERP...MES doesn't always break loudly. Sometimes it just... struggles quietly.

What That Looks Like on the Floor

Here's how that disconnect plays out:

- A work order is missing key details → MES can't process it correctly
- Inventory is misallocated → lots can't move forward
- A BOM (Bill of Material) is incorrect → the system expects materials that don't exist
- Planning data is off → production flow feels "off," but not obviously broken

And if you don't understand ERP, you stay stuck in MES trying to fix something that didn't start there.

The Hard Lesson

I learned this the long way, spending extended time troubleshooting:

- Rechecking configurations
- Restarting processes
- Escalating issues

Only to find out later:

"This came from ERP."

That's a frustrating place to be—because you realize the problem was never yours to fix the way you were trying to fix it. You were just the one closest to the impact.

Why This Matters Early (Not Later)

Looking back, I wish I had understood this sooner: **MES and ERP are not separate worlds—they are dependent systems.**

- ERP plans the work
- MES executes the work
- ERP defines the materials
- MES tracks their usage
- ERP creates the order
- MES moves it through production

If one is off, the other reflects it.

What I Would Do Differently

If I were starting over, I wouldn't wait until something breaks to learn ERP. I would:

- Ask how work orders are created
- Understand where inventory data comes from
- Learn how BOMs are structured
- Pay attention to how planning decisions show up in MES

Not to become an ERP expert—but to understand the flow of information before it reaches me.

The Bigger Shift in Thinking

This was the shift for me: I stopped seeing MES as the *source* of truth. And started seeing it as part of a larger system pipeline. That changed how I troubleshoot, how I ask questions, and how quickly I get to real answers.

What No One Tells You

A lot of early MES frustration comes from this gap. You think: "I just need to learn MES better." But sometimes the real answer is: "I need to understand what's feeding MES."

Takeaway

If something doesn't make sense in MES, ask yourself: **"Did this start here—or did it start upstream?"** That one question can save you hours.

Understanding ERP: The System No One Told You About

When people first learn MES, they're usually trained on:

- How to move lots
- How to follow routes
- How to read the system

What they're *not* taught—at least not clearly—is where all of that information comes from.

That's where ERP comes in.

So, What is ERP? (Plain Language)

ERP stands for **Enterprise Resource Planning**. ERP is the system that plans and organizes everything the business needs to run production. If MES is focused on *execution*, ERP is focused on *planning*.

Think of It Like This

Imagine you're building a house.

- ERP is the system that:
  - Decides how many houses to build

    - Orders the materials
    - Creates the construction plan
    - Schedules the work
- MES is the system that:
    - Tracks the actual building process
    - Records each step as it happens
    - Makes sure the work follows the plan

One plans. One executes. You need both.

What ERP Actually Does

Behind the scenes, ERP is responsible for things like:

- **Work Orders**

It creates the instructions for what needs to be built.

- **Inventory Management**

It tracks raw materials, parts, and finished goods.

- **Bill of Materials (BOM)**

It defines what components are needed to make a product.

- **Production Planning**

It decides *when* and *how much* to produce.

- **Supply Chain Coordination**

It ensures materials are available when needed.

Before anything shows up in MES, it usually starts here.

Chapter 3

How ERP and MES Work Together

This is where things finally click. ERP sends instructions. MES carries them out.

Here's a simplified flow:

1. ERP creates a work order
2. ERP defines the materials and requirements
3. That information is sent to MES
4. MES uses that data to guide production
5. MES tracks everything happening in real time
6. MES sends updates back to ERP

So the communication goes both ways:

- ERP → MES (planning → execution)
- MES → ERP (status → reporting)

Why This Matters for You as a Technician

Here's the part that most training skips—and where people struggle. When something goes wrong in MES, it doesn't always mean MES is the problem. Because MES is often working with what it was given.

If ERP sends:

- Incorrect inventory
- Incomplete work orders
- Wrong BOM structures

Then MES will reflect those problems—not always with a clear error message, but with behavior that doesn't make sense.

What This Looks Like in Real Life

You might see:

- A lot that won't move forward
- A step that won't complete
- Materials that the system says are missing
- Data that looks "off," but not clearly broken

You check everything in MES… and it still doesn't add up. That's your signal:

The issue may have started in ERP.

Why No One Explains This Early (But Should)

Most MES training is designed to get you functional quickly. So it focuses on:

- Buttons
- Screens
- Procedures

But not the full system flow. The problem is, without understanding ERP:

- You troubleshoot in the wrong place
- You waste time chasing symptoms
- You miss the root cause

How to Start Thinking Differently

Instead of asking: “What’s wrong with MES?” Start asking: “Where did this data come from?” That one shift changes everything.

What You Should Learn (Without Going Too Deep)

You don’t need to become an ERP expert. But you *should* understand:

- Where work orders originate
- How inventory is defined and tracked
- What a BOM is and why it matters
- How planning decisions show up in MES

Just knowing this puts you ahead of most beginners.

Bottom Line

ERP plans the work. MES executes the work.

If you only understand MES, you’re seeing half the picture. And in this job, half the picture isn’t enough.

Understanding BOM: The Blueprint Behind Every Product

Before a single lot moves in MES…Before a work order even shows up…There's something quietly defining *everything* that needs to happen. That's the BOM.

What is a BOM? (Plain Language)

BOM stands for **Bill of Materials**.

Here's the simplest way to think about it:

A BOM is the complete list of everything required to build a product.

Not just the main parts—*everything*:

- Raw materials
- Components
- Sub-assemblies
- Quantities of each
- Sometimes even the order they're used

Think of It Like a Recipe

If you've ever followed a recipe, you already understand BOM. Let's say you're making a cake. The recipe tells you:

- Ingredients (flour, sugar, eggs)
- How much of each
- What's needed before you even start

That's your BOM. Now imagine trying to bake without it:

- You don't know what's required
- You don't know how much to use
- You don't know if you're missing something

That's what production looks like without a proper BOM.

Where BOM Lives (And Why That Matters)

The BOM is created and managed in the **ERP system**. That's important. Because by the time MES sees anything:

- The BOM is already defined
- The materials are already expected
- The structure is already set

MES doesn't decide what's needed.

It trusts what ERP tells it.

How BOM Shows Up in MES

You won't always see the full BOM directly in MES.

But you *will* see its effects:

- The system expects certain materials at certain steps
- It checks whether required components are available
- It may block progress if something is missing

- It tracks how materials are used during production

So even if you don't realize it, you're working with the BOM every day.

What Happens When a BOM is Wrong

This is where things get real.

If the BOM is incorrect in ERP, MES doesn't always throw a clear error.

Instead, you get situations like:

- The system says material is required—but it doesn't exist
- A lot can't move forward due to "missing components"
- Quantities don't line up with what's actually available
- Production gets stuck in ways that don't make sense

And if you don't know about BOMs, you'll stay stuck in MES trying to fix something that started somewhere else.

Why BOM Matters to You as a Technician

You're not responsible for creating BOMs. But you *are* affected by them.

Understanding BOM helps you:

- Recognize when an issue isn't MES-related
- Ask better questions when troubleshooting
- Communicate more effectively with engineers and planners
- Avoid wasting time chasing the wrong problem

The Key Shift

Instead of thinking: "Why is MES asking for this?"

Start thinking: "What does the BOM require here?"

That question alone will make you faster and more accurate.

Quick Reference Sheet: ERP vs MES vs BOM

This is your "keep it simple under pressure" guide.

ERP (Enterprise Resource Planning)

Role: Plans the work

- Creates work orders
- Defines inventory
- Builds and manages BOMs
- Schedules production
- Manages supply and demand

➢ *Think:* "Where everything starts"

BOM (Bill of Materials)

Role: Defines what's needed

- Lists all required materials
- Specifies quantities
- Structures how products are built
- Lives inside ERP
- ➢ *Think:* "The blueprint or recipe"

MES (Manufacturing Execution System)

Role: Executes the work

- Moves lots through production
- Tracks progress in real time
- Enforces process steps
- Validates materials and actions
- Records what actually happened
- ➢ *Think:* "Where the work happens"

How They Work Together (Simple Flow)

1. **ERP** creates the work order
2. **BOM (inside ERP)** defines required materials
3. ERP sends that information to MES
4. **MES** executes and tracks production

5. MES sends status updates back to ERP

⚠ Troubleshooting Shortcut

When something goes wrong, ask:

- Is this a **planning issue**? → Check ERP
- Is this a **material definition issue**? → Check BOM
- Is this an **execution issue**? → Check MES

Bottom Line

- ERP decides *what* to build
- BOM defines *what's needed* to build it
- MES handles *how it gets built*

Miss one of these, and the whole system feels confusing. Understand all three, and things start to make sense fast.

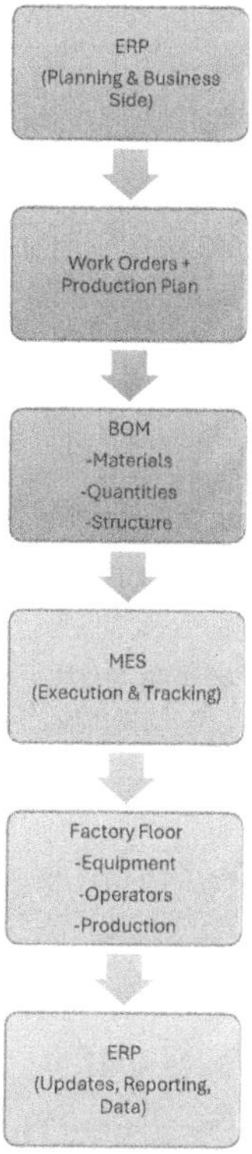

Real Troubleshooting Walkthrough: When It's NOT MES

Scenario: "The Lot Won't Move"

You're on shift. An operator calls you over: "This lot won't move to the next step."

Simple enough. Happens all the time.

Step 1: You Check MES (Like You Were Trained)

You look at:

- ✓ Current operation
- ✓ Route
- ✓ System status
- ✓ No clear error messages

Everything looks fine. But the lot still won't move.

Step 2: You Try Basic Fixes

- Refresh the system
- Retry the transaction
- Double-check permissions

Still stuck. Now production is waiting.

Step 3: Something Feels Off

This is where experience starts to kick in. There's no hard error—but something isn't right. So instead of asking: "What's wrong with MES?" You ask: **"What is MES expecting right now?"**

Step 4: Follow the Requirement (This Leads You to BOM)

You dig a little deeper and notice: The system is expecting a material that isn't available.

But no one mentioned missing materials. Now you're not just looking at MES anymore. You're thinking about the BOM.

Step 5: Trace It Upstream (ERP)

You escalate or investigate further and find:

- The BOM requires **Component A**
- But inventory shows **Component A isn't available**
- Or worse—it was never properly allocated to the work order

Now it clicks.

Root Cause

The issue didn't start in MES.

✓ MES was enforcing the requirement
✓ The BOM defined the requirement

✗ But ERP had incorrect or incomplete data

What MES Was Actually Doing

MES wasn't broken. It was doing exactly what it's supposed to do: **Blocking production because required materials weren't satisfied.**

Why This Confuses New Technicians

Because there's no obvious error that says: "Hey—this is an ERP issue."

Instead, you get:

- A stuck lot
- No clear explanation

- Pressure to fix it quickly

And if you don't understand ERP + BOM, you stay stuck in the wrong layer.

What an Experienced Technician Does Differently

They don't just check MES.

They think in layers:

1. **Is MES executing correctly?**
2. **What is MES expecting (BOM)?**
3. **Where did that expectation come from (ERP)?**

That's how you move faster.

The Real Skill

Anyone can follow steps in MES. The real skill is: **Knowing where the problem started—even if it didn't start with you.**

When something breaks:

- **ERP Issue → Bad plan**
- **BOM Issue → Wrong requirements**
- **MES Issue → Bad execution**

The system didn't fail you. It showed you exactly what it was told to do. Your job is to figure out who told it wrong.

Part 3
Creating Processes and Production

Chapter 4

Building the Process: Operations, Recipes, and Tools

At some point, you move from using MES… to helping define how it works. This is where you start:

- Creating operations
- Defining recipes
- Assigning tools
- Structuring how production flows

And this is where small decisions can turn into big problems later.

What You're Really Doing (Whether You Realize It or Not)

When you create: An operation, recipe, or a tool configuration, you're not just entering data. You are building the logic that production will follow.

Every operator, technician, and engineer will rely on what you set up. So if it's unclear, inconsistent, or poorly named, that confusion spreads fast.

Operations: Defining the Flow

What is an Operation? An operation is a **step in the manufacturing process**.

It answers: “What happens at this stage of production?”

Examples:

- Inspection
- Etch
- Deposition
- Assembly
- Test

Why Operations Matter

Operations define:

- The sequence of production
- Where a lot goes next
- What needs to be done before moving forward

If operations are wrong:

- Lots get routed incorrectly
- Steps get skipped or repeated
- Production flow breaks down

Real-World Insight

Early on, it’s easy to think: “I’ll just name it something quick so we can move on.” That works… until someone else has to interpret it. Or until *you* come back months later and don’t remember what you meant.

Recipes: Defining How the Work Gets Done

What is a Recipe?

A recipe is the **set of instructions the equipment follows** to perform an operation.

It can include:

- Process parameters
- Timing
- Temperature
- Tool-specific instructions

Simple Way to Think About It

- **Operation = What step is happening**
- **Recipe = How that step is performed**

Why Recipes Matter

If a recipe is wrong:

- The process may fail
- Product quality can be affected
- You may not catch the issue immediately

And in some environments, that gets expensive fast.

Tools: Where Execution Happens

What is a Tool?

A tool is the **equipment used to perform the operation**.

Examples:

- Machines
- Chambers
- Workstations

Why Tool Setup Matters

MES uses tool definitions to:

- Route work correctly
- Ensure compatibility with recipes
- Track usage and performance

If tools are set up incorrectly:

- The wrong equipment may be used
- Recipes may not match
- Data tracking becomes unreliable

The Part Most People Underestimate: Naming Conventions

This is where things quietly go wrong in a lot of environments.

Why Naming Matters More Than You Think

Names are how people understand the system. Not just today—but months or years from now. If your naming is:

- Vague
- Inconsistent
- Too abbreviated
- Or overly complex

Then people will:

- Misinterpret processes
- Use the wrong recipes
- Waste time figuring things out

❌ Bad Naming Examples

- Op1
- Test_New
- RecipeA
- Final_v2_fixed

These might make sense in the moment—but not long-term.

Good Naming Principles

A good name should tell you:

- What the step is
- What it's used for
- Sometimes where or how it applies

Without needing extra explanation.

Stronger Examples

Instead of:

- Op1

Use:

- Incoming_Inspection

Instead of:

- RecipeA

Use:

- Etch_Silicon_Standard_v1

Instead of:

- Test_New

Use:

- Final_Test_Electrical_v2

What Engineers and Teams Need

Remember—MES is not a solo environment. Engineers, planners, and other technicians rely on what you name things. Good naming helps:

- Engineers quickly identify processes
- Teams troubleshoot faster
- New hires ramp up quicker

- Everyone stay aligned

The Long-Term Impact

Here's what people don't think about: You might create something in 10 minutes…But people may interact with it for years. So ask yourself: "Will this still make sense to someone who wasn't here when I created it?"

⚠ Common Mistakes to Avoid

- Naming based on memory instead of clarity
- Using inconsistent formats across similar processes
- Overusing abbreviations that only you understand
- Renaming things too often (creates confusion and tracking issues)

A Simple Naming Framework You Can Use

Keep it structured:

[Process] + [Material/Product] + [Type/Variation] + [Version]

Example:

- Deposition_Aluminum_Standard_v1
- Inspection_Wafer_Visual_v2
- Assembly_Module_Final_v1

This keeps things:

- Clear

- Searchable
- Consistent

What No One Tells You

You don't realize how important naming is…

Until:

- Something breaks
- Someone asks questions
- Or you have to trace an issue back through the system

And suddenly:

- Names are all you have to go on

Takeaway

- Operations define *what happens*
- Recipes define *how it happens*
- Tools define *where it happens*
- Naming defines whether anyone can understand it later

In MES, clarity isn't optional.

The way you build and name things today becomes the way others understand the system tomorrow.

Step-by-Step Guide: How to Create an Operation and Recipe Correctly

Before we get into steps, understand this:

You're not just creating a step in the system. You're defining how production will run.

So don't rush this.

⚠ Before You Touch the System (Most People Skip This)

Take a minute and ask:

- What is this step supposed to accomplish?
- Where does it sit in the process flow?
- What happens before and after it?
- What tool will be used?
- What recipe is required?

If you can't answer these clearly, don't build yet.

Clarity first. Configuration second.

PART 1: Creating an Operation

Step 1: Define the Purpose of the Operation

Ask: "What is happening at this step?"

Be specific. Not just:

- "Process"

But:

- “Wafer Cleaning”
- “Final Inspection”
- “Oxide Etch”

➢ If the purpose isn’t clear, everything after this becomes guesswork.

Step 2: Place It in the Correct Sequence

Operations don’t exist alone—they are part of a flow.

Ask:

- What operation comes before this?
- What operation comes after this?

➢ A misplaced operation can break the entire routing of a lot.

Step 3: Assign the Correct Work Center / Area

This defines *where* the work happens.

- Which department?
- Which tool group?
- Which production area?

➢ This affects routing, visibility, and reporting.

Step 4: Link the Appropriate Tool(s)

Now you define *what equipment* can perform this operation.

Ask:

- Which tools are capable?
- Are there multiple valid tools?
- Are there restrictions?

➢ If tools are wrong or missing, production will stall.

Step 5: Apply a Clear Naming Convention

Use something that explains the step without extra context.

Good format:

[Process]_[Material/Stage]_[Type]

Example:

- Etch_Silicon_Main
- Inspection_Final_Visual

➢ If someone has to ask what it means, the name failed.

Step 6: Add Required Attributes (If Applicable)

Depending on your system, this may include:

- Hold points

- Required validations
- Data collection steps
- Operator actions
- ➢ This is where control and quality checks live.

Checkpoint Before Moving On

Before you save or release the operation, ask:

- Does this step make sense in the full process flow?
- Would someone new understand what this is?
- Is it connected to the correct tools?

If not—fix it now.

PART 2: Creating a Recipe

Step 1: Understand the Process Requirements

Before creating a recipe, you need to know:

- What is this process doing?
- What parameters matter?
- What outcome is expected?
- ➢ You're defining behavior—not just filling in fields.

Step 2: Match the Recipe to the Operation

Every recipe should clearly align with its operation.

Ask:

- Is this the correct recipe for this step?
- Are there variations (standard vs custom)?

➢ Misalignment here creates confusion and risk.

Step 3: Define Key Parameters

This may include:

- Temperature
- Time
- Pressure
- Speed
- Tool-specific settings

➢ Even if engineers define these, you need to understand what you're entering.

Step 4: Link the Recipe to the Correct Tool(s)

Not all recipes work on all tools.

Ask:

- Is this recipe compatible with the assigned tool?
- Are there tool-specific versions?

➢ This prevents execution failures.

Step 5: Use a Clear Naming Convention

Avoid vague names.

Good format:

[Process]_[Material]_[Type]_[Version]

Examples:

- Etch_Silicon_Standard_v1
- Deposition_Aluminum_HighTemp_v2

➢ Versioning matters. It helps track changes and avoid confusion.

Step 6: Validate Before Release

Before making the recipe active:

- Has it been reviewed or approved (if required)?
- Does it match engineering specs?
- Has it been tested or verified?

➢ A bad recipe can impact production quality—not just flow.

PART 3: Linking It All Together

Now bring it all into alignment:

- Operation → Defines the step

- Recipe → Defines how it runs
- Tool → Executes it

Ask yourself:

"If a lot reaches this step, will everything it needs be available and correct?"

If the answer isn't clearly yes—something is missing.

⚠️ Common Mistakes (And How to Avoid Them)

❌ **Rushing the setup**

- Leads to rework and confusion

❌ **Poor naming**

- Slows down everyone later

❌ **Not verifying tool compatibility**

- Causes execution failures

❌ **Skipping validation**

- Creates hidden problems

How Strong Technicians Approach This

They don't just "set it up."

They think:

- How will this be used?

- Who else needs to understand this?
- What could go wrong here?

That's the difference.

Quick Build Checklist (Use This Every Time)

Before finalizing:

✓ Operation purpose is clear
✓ Correct sequence in flow
✓ Tools are properly assigned
✓ Recipe matches operation
✓ Naming is clear and consistent
✓ Parameters are verified
✓ Setup has been reviewed/tested

Closing

Anyone can create an operation or recipe.

Not everyone creates one that works, makes sense, and lasts.

That's the standard you should aim for.

Chapter 5

What "Configuration" Really Means in MES

If you spend any time around MES, you're going to hear this word a lot:

"Check the configuration."
"It's probably a config issue."
"We need to update the configuration."

And in the beginning, that can feel vague and frustrating. So let's make it clear.

What is Configuration? (Plain Language)

Configuration is how the system is set up to behave. Not how it *could* work, not how it *should* work. How it is currently defined to work.

A Simple Way to Think About It

If MES were a machine:

- **Configuration = the settings, rules, and instructions that control it**

You're not changing the machine itself. You're changing:

- What it does
- How it responds
- What it allows or blocks

What Falls Under Configuration in MES

Configuration includes things like:

- Operations (process steps)
- Recipes (how steps are performed)
- Tool assignments
- Routing logic
- Validation rules
- Data requirements
- Permissions and restrictions

Everything you've been reading about?

That's configuration.

Why Configuration Matters So Much

Here's the part that matters on the job: MES doesn't "think." It follows configuration.

So if something is happening in the system:

- A lot won't move
- A step is blocked
- A material is required

It's not random. It's the system following the rules it was given.

What This Means for You

When you hear:

“Check the configuration”

What someone is really saying is:

“Check how the system is set up to behave in this situation.”

Real-World Example

A lot can’t move to the next operation. You might assume:

- System glitch
- Tool issue
- User error

But when you check the configuration, you find:

- The operation requires a specific recipe
- That recipe is only allowed on certain tools
- The current tool isn’t included

So MES blocks the move.

Not because it’s broken but because it’s configured that way.

Configuration vs. Error (Important Difference)

This is where new technicians get tripped up.

❌ **Thinking:**

“The system is wrong.”

Reality:

"The system is doing exactly what it was configured to do."

Two Types of Configuration Issues

1. Incorrect Configuration

- Wrong tool assigned
- Missing recipe
- Bad routing
- Incorrect rules

➢ The system behaves incorrectly because it was set up incorrectly.

2. Missing Configuration

- Operation not fully defined
- Recipe not linked
- Required data not configured

➢ The system can't proceed because it doesn't have what it needs.

Why This Word Comes Up So Often

Because most MES issues come back to this: Configuration controls behavior. So when something goes wrong, people naturally look there first.

The Shift That Changes Everything

Early on, you react to problems. Later, you start thinking like this:

"What configuration is driving this behavior?" That's when you stop guessing and start understanding.

How This Connects to Everything You've Learned

- **ERP** → Defines what should happen
- **BOM** → Defines what's needed
- **Configuration (in MES)** → Defines how it actually plays out

What Strong Technicians Know

They don't just use MES.

They understand:

- How it's built
- How it's configured
- And how those configurations drive outcomes

Quick Reference: Configuration

Configuration = System Rules + Setup + Behavior

If something in MES:

- Doesn't make sense

- Isn't working
- Or is blocking progress

➤ Look at the configuration.

Closing:

MES doesn't make decisions. It follows configuration. If you understand the configuration, you understand the system.

➤ **Configuration Breakdown Scenario: One Small Setup, Big Impact**

Scenario: "Everything Looks Right… But Nothing Works"

You're on shift, and production is moving—until it isn't. An operator flags you down: "This lot won't process. It keeps getting blocked." You've seen this before. Should be quick.

Step 1: Check MES (Surface Level)

You pull up the lot and review:

✓ Current operation

✓ Route

✓ Tool availability

✓ No obvious system errors

At first glance, everything looks fine. But the lot still won't move.

Step 2: Look Closer at the Error

You dig a little deeper and finally see it: "Invalid recipe for selected tool" Now you have something.

Step 3: Check the Recipe

You verify:

- ✓ The recipe exists
- ✓ It's active
- ✓ It's the correct one for the operation

Still doesn't explain the issue.

Step 4: Check the Tool Configuration

Now you shift your focus. You look at the tool setup and find:

- ➢ The recipe is **not linked to this specific tool**

What This Means

Even though:

- The operation is correct
- The recipe is correct
- The tool is available

The configuration does not allow them to work together. So MES blocks the process.

Step 5: Trace It Back Further (Why Was It Set Up This Way?)

Now you ask a better question:

"Why isn't this recipe linked to the tool?"

Possible reasons:

- It was never added during setup
- It was removed during an update
- A new tool was added but not fully configured
- Version mismatch between recipe and tool

This is where configuration mistakes show up.

Where ERP and BOM Fit Into This

At first, this looked like a simple MES issue. But zoom out:

- **ERP** created the work order expecting this process to happen
- **BOM** defined that this step requires this type of process
- **MES configuration** failed to properly connect:
  - Operation
  - Recipe
  - Tool

So even though planning was correct…

Execution failed due to configuration.

The Chain Reaction

One missing link caused:

- Lot blocked
- Operator waiting
- Production delay
- Escalation to technicians

All from something that took seconds to miss during setup.

The Fix

Once identified:

- Link the correct recipe to the tool
- Validate compatibility
- Retest the lot movement

Now everything works.

Why This Matters

This wasn't:

- A system failure
- A broken tool
- A bad work order

It was a **configuration gap**. And those are some of the hardest issues to spot—because everything *looks* correct at a glance.

What a New Technician Might Do

- Restart the system

- Retry the process multiple times
- Escalate without clear direction

What a Strong Technician Does

They think in layers:

1. What is MES doing?
2. What is MES expecting?
3. How is the system configured to handle this?

That's how they find the issue faster.

Key Lesson

If MES is blocking something, it's usually doing it for a reason.

Your job is to understand the rule it's enforcing.

⚠ Where This Goes Wrong in Real Environments

- Configuration is rushed
- Changes aren't fully tested
- Naming is unclear
- Documentation is missing

And over time, small gaps turn into bigger issues.

Quick Breakdown

Layer	Role	Status in This Scenario
ERP	Planned the work	✔ Correct
BOM	Defined requirements	✔ Correct
MES Config	Controlled execution	✖ Incomplete

The system didn't break. It followed the setup it was given—exactly. And one missing configuration was enough to stop everything.

Part 4
A Day in the Life of an MES Technician

Chapter 6
Office vs Shop Floolr Reality

Two Versions of the Same Job

When most people think of MES, they imagine someone sitting at a computer:

- Building routes
- Creating operations
- Defining recipes
- Managing configurations

That was my beginning. I learned MES from the office side first. What I didn't fully understand at the time was that MES doesn't just live on a screen. It lives on the Shop Floolr. And those are two very different experiences of the same system.

Part 1: MES in the Office (The System Side)

This is where most MES technicians start. Your day might look like:

- Creating or modifying operations
- Building or updating recipes
- Configuring tools and tool groups
- Adjusting routing logic
- Supporting engineers with system setup

- Reviewing errors or change requests

From this perspective, MES feels:

- Structured
- Controlled
- Logical
- Clean

You are working with:

- Data
- Rules
- Configurations
- System behavior

And most of the time, nothing is physically breaking in front of you. So it feels like you're managing the system.

The Hidden Limitation of the Office View

The problem is: In the office, everything works *in theory*. But MES doesn't live in theory.

It lives in:

- Real equipment
- Real operators
- Real timing issues
- Real production pressure

And none of that fully shows up on a screen.

Part 2: MES on the Shop Floolr (The Reality Side)

This is where everything becomes real. On the Shop Floolr, MES is not abstract. It is visible in action. You see:

- Operators trying to move lots through steps
- Equipment waiting for the correct recipe or setup
- Production lines slowing down or stopping
- People calling out errors that don't make sense in the system

This is where MES is *experienced*, not just configured.

What MES Looks Like on the Floor

Instead of:

- "Operation created successfully"

You see:

- A lot stuck at a station

Instead of:

- "Recipe linked to tool"

You see:

- A machine refusing to run the process

Instead of:

- "Configuration updated"

You see:

- Production waiting with no clear explanation

The Emotional Difference

In the office, MES feels like: "I'm building the system." On the floor, MES feels like: "Why isn't this working right now?" That gap matters.

The Disconnect Most Technicians Don't Get Taught

When I first started, I didn't spend time on the Shop Floolr. So I understood:

- How to build routes
- How to configure operations
- How recipes should behave

But I didn't understand:

- How operators actually interact with MES
- How delays show up in real time
- How equipment behavior affects system flow
- How small configuration gaps become production stops

That missing perspective cost me time in troubleshooting.

What Changes When You Understand Both Sides

Once you've seen MES on the floor, your thinking changes. You stop asking: "What is wrong in the system?" And start asking: "What is happening physically that the system is reacting to?"

Example Shift in Thinking

❌ **Office-only mindset:** "The operation is configured correctly."

✓ **Full perspective mindset:**

"The operation is correct in the system—but what is the tool actually doing right now?"

How MES Really Works in Practice

MES is constantly translating between:

Office side:

- Planning
- Configuration
- Structure

Shop Floolr side:

- Physical execution
- Human interaction
- Machine behavior

And when those two don't align perfectly…That's where issues show up.

What You Are Actually Doing as an MES Technician

You are not just:

- Building configurations
- Fixing system errors
- Creating recipes

You are also:

- Interpreting what the floor is experiencing
- Translating physical issues into system logic
- Bridging communication between engineers, operators, and data

Why Shop Floolr Understanding Matters

Without Shop Floolr exposure, you might:

- Fix the wrong layer of the problem
- Miss timing or sequencing issues
- Overlook operator-driven behavior
- Assume system logic is always the root cause

But with Shop Floolr understanding, you can see:

- Where delays actually begin
- How MES decisions show up physically
- What "system errors" look like in real production

Simple Comparison

Area	What You See	What It Feels Like
Office MES	Routes, recipes, configs	Structured, controlled
Shop Floolr MES	Lots, machines, operators	Real-time pressure, visible impact

Key Lesson

MES is not just built in the office. It is proven on the Shop Floolr. If you only understand one side, you're working with half the system.

Closing Reflection

I started my career only seeing MES from behind a desk. I understood how to build the system, but I didn't yet understand how the system lived in production. Once I learned both sides, troubleshooting became faster, clearer, and more accurate. Because I stopped thinking in terms of:

"What did I configure?"

And started thinking in terms of:

"What is actually happening on the floor?"

What an MES Technician Actually Fixes on the Shop Floolr

Before we talk about failures or troubleshooting frameworks, it helps to ground this in something real.

On the Shop Floolr, MES issues don't always look like "system problems." They show up as **production problems that just happen to involve the system.**

A common example:

Example: A Lot Stuck at a Tool

An operator calls over:

"This lot won't move. It's stuck at this tool step."

At first glance, it looks simple. But when you arrive, you might find:

- ✓ The tool is available
- ✓ The operator is ready
- ✓ The lot is valid
- ✓ No obvious MES error message

But the lot still won't proceed.

What You Might Actually Be Fixing

After digging into MES, you might discover things like:

- The **recipe is not properly linked to the tool configuration**

- The operation requires a **specific version of a recipe that isn't active**
- The tool is missing a **required configuration parameter**
- The lot is waiting on a **system validation step tied to BOM or ERP data**

So what looked like:

"a stuck lot"

Is actually:

a chain of configuration, recipe, and system logic dependencies not fully aligned

Why This Matters

From the floor, it looks like a machine issue.

From MES, it looks like a data/configuration issue.

From reality, it's usually:

a mismatch between how the system was configured and how the process is actually being executed

Key Insight

Most MES work on the floor is not "fixing broken systems."

It's:

figuring out why the system is behaving exactly the way it was configured to behave—even if that behavior is blocking production

Chapter 7

How MES Failures Actually Happen

The First Misunderstanding

When something breaks in MES, the natural reaction is: "Something is wrong with the system." But in most environments, that's not accurate. MES rarely "fails" randomly.

Instead, it is usually: executing correctly based on incomplete, incorrect, or misaligned setup data

Failure Point #1: Upstream Data (ERP & Planning)

A lot of MES "failures" actually start before MES ever sees the work.

Examples:

- Work orders created incorrectly
- Missing or inaccurate BOM data
- Inventory not properly allocated
- Wrong production quantities or timing

MES receives this data and executes it exactly as provided.

So when things don't make sense on the floor…

The root cause often started upstream.

Failure Point #2: Configuration Gaps

This is the biggest hidden issue.

MES depends heavily on configuration:

- Operations
- Recipes
- Tool assignments
- Routing rules

If even one link is missing or misaligned:

- Lots get stuck
- Steps are blocked
- Tools reject processing

Not because MES is broken…

But because MES is enforcing its rules.

Failure Point #3: Version Mismatches

Another common issue:

- Recipe updated but not linked correctly
- Operation version changed but not fully deployed
- Tool config not aligned with latest process

Result:

System and process are no longer speaking the same language

Failure Point #4: Human Assumptions

Sometimes the system is correct—but assumptions are not.

Examples:

- "This tool should work with this recipe" (but isn't configured that way)
- "This operation is standard" (but has exceptions built in)
- "This lot should move" (but is missing required conditions)

MES doesn't assume. It verifies.

Key Truth

MES rarely breaks. It reveals mismatches.

How to Troubleshoot Like a Senior MES Technician (Framework-Based)

The Shift That Changes Everything

Junior technicians ask:

"What's wrong with MES?"

Senior technicians ask:

"What condition is MES enforcing right now?"

That difference is everything.

The 5-Layer Troubleshooting Framework

When something goes wrong, work through these layers in order:

Layer 1: What is happening physically? (Shop Floolr Reality)

- Is the tool running?
- Is the operator able to proceed?
- Is the lot actually moving or stuck?

➢ Start with reality, not the system.

Layer 2: What is MES showing?

- Status of lot
- Current operation
- Error messages (if any)
- Hold reasons

➢ This tells you what MES believes is happening.

Layer 3: What is MES expecting?

This is where most real issues are found.

Ask:

- What recipe is required?
- What tool is required?
- What validation must pass?

- What data must exist?

➢ This reveals the rule being enforced.

Layer 4: Where does that expectation come from? (ERP / BOM)

Now zoom out:

- Is the work order correct?
- Is the BOM accurate?
- Was the data properly released?

➢ This identifies upstream causes.

Layer 5: Is configuration correct?

Now check:

- Operation setup
- Recipe linking
- Tool assignment
- Routing logic

➢ This determines whether MES is correctly built.

How Seniors Think Differently

Instead of reacting, they trace:

Floor → MES behavior → MES expectation → Source of expectation → Configuration

They don't jump steps.

They follow the system logic.

⚠ Common Mistake (Junior Approach)

- Restart system
- Retry action repeatedly
- Escalate without tracing layers
- Assume "system issue" too early

✓ Senior Approach

- Observe physical issue
- Confirm MES behavior
- Identify rule being enforced
- Trace source (ERP/BOM/config)
- Fix root cause—not symptom

Final Insight

MES troubleshooting is not about finding what is broken. It is about finding where reality and configuration no longer match.

Chapter 8
On-Call Shift Story: When Everything Breaks at Once

It's late in the shift. Production is running—until it isn't. Your phone rings. "Multiple lots are stuck across different tools. We can't move anything forward." That's never a good sentence to hear. You walk into a situation where:

- Operators are waiting
- Tools are idle
- Supervisors are asking for updates
- Everyone assumes "MES is down"

But nothing is actually down.

Step 1: First Look at the Floor

You go to the Shop Floolr. What you see:

- Lots sitting at different operations
- Tools showing available
- Operators unsure what to do next
- No physical equipment failure

This is important. Because physically, everything looks… normal.

Step 2: Check MES Status

You pull up the system:

✓ Lots are valid
✓ Routes are correct
✓ No system outage
✓ No obvious alarms

So why is nothing moving? Now you stop thinking in terms of "system failure." And start thinking in terms of: "What rule is stopping movement across multiple lots?"

Step 3: Pattern Recognition Begins

You notice something: All stuck lots share one thing.

- Same operation step
- Same recipe version
- Same tool group requirement

This is not random. This is a **systematic block**.

Step 4: Trace the Expectation (MES Logic)

Now you check what MES is enforcing:

- Required recipe version ✔ mismatch
- Tool qualification ✔ incomplete
- Configuration rule ✔ recently updated

Now the picture forms. Nothing is broken. **MES is enforcing a rule that is now too restrictive or incomplete.**

Step 5: Find the Root Cause

You trace upstream:

- A recipe update was deployed earlier
- Tool linking wasn't fully completed
- One configuration rule wasn't updated across all tool groups

So now: MES is correctly blocking movement based on incomplete configuration.

Step 6: Fix and Validate

- Correct tool-recipe linking
- Update configuration rules
- Validate affected lots
- Release hold conditions

Gradually:

- Lots begin moving again
- Tools resume operation
- Production flow stabilizes

Key Takeaway from the Shift

Nothing was "down." What happened was: a small configuration gap created a system-wide production blockage.

And because MES is interconnected, one missing piece echoed across multiple lots.

What You Learn From Nights Like This

- Problems rarely appear where they start
- Multiple failures often trace back to one configuration change
- The system is consistent—even when it feels confusing
- Your job is pattern recognition, not panic reaction

⚠ Top 10 MES Issues (And What They Actually Mean)

This is your "decode the language of MES" section.

1. "Lot is stuck at operation"

➢ Usually: missing recipe, tool mismatch, or unmet configuration rule

2. "Invalid recipe error"

➢ 👉 Recipe not properly linked to tool or wrong version active

3. "Tool unavailable"

➢ Tool may be free physically, but not qualified in MES configuration

4. “Hold placed on lot”

➢ Often upstream data issue (ERP, BOM, or quality rule trigger)

5. “Cannot move to next step”

➢ Missing validation requirement or routing rule not satisfied

6. “Material not found”

➢ BOM mismatch or inventory not properly allocated in ERP

7. “System timeout or delay”

➢ Interface or integration lag—not always MES itself

8. “Incorrect operation sequence”

➢ Routing configuration error or version mismatch in process flow

9. “Tool not eligible for process”

➢ Tool not properly configured or recipe-tool association missing

10. “Data mismatch error”

➢ ERP data, BOM structure, or MES configuration not aligned

What These Really Teach You

Most MES errors fall into 4 categories:

- **Configuration issues**

- **Data mismatches (ERP/BOM)**
- **Tool/recipe alignment problems**
- **Process expectation gaps**

Not system failure. Not random breakdowns.

Final Insight for This Section

MES doesn't fail randomly. It exposes misalignment between planning, configuration, and execution. Your job is not just to fix errors—but to understand what rule is being enforced and why.

The MES Technician Mindset: How to Think in Layers, Not Tickets

The Biggest Shift in This Role

When most people start in MES, they are trained to think in tickets:

- "Here is the problem"
- "Here is the error"
- "Here is the fix"

So their thinking becomes:

Find issue → Fix issue → Close ticket

That works for simple problems.

But MES problems are rarely simple. Because what you see on the screen is often **not where the problem started**.

Why Ticket Thinking Fails in MES

A ticket usually shows:

- A symptom
- A moment in time
- A surface-level error

But MES is a layered system:

- ERP defines the work
- BOM defines the materials
- Configuration defines behavior
- MES executes the logic
- The Shop Floolr reveals reality

So if you only focus on the ticket, you are usually fixing:

the *last visible layer*, not the root cause

Layered Thinking: The Core Skill of Strong MES Technicians

Instead of thinking in tickets, you start thinking in layers. Every issue has depth.

Layer 1: Physical Reality (Shop Floolr)

Start here every time:

- What is actually happening at the tool?
- Is production moving or stopped?
- What is the operator experiencing?

- This is the truth layer. No system interpretation yet.

Layer 2: MES Behavior (What the System Shows)

Now check MES:

- Lot status
- Operation step
- Errors or holds
- Movement restrictions
- This is MES's interpretation of reality.

Layer 3: MES Expectation (What It Wants to Happen)

Now shift your thinking: What is MES expecting right now?

- Required recipe
- Tool qualification
- Validation rules
- Process conditions
- This is where most issues are actually discovered.

Layer 4: Configuration (How the System Was Built)

Now zoom into structure:

- Operations setup
- Recipe linking

- Tool assignments
- Routing logic

➢ This is how MES was instructed to behave.

Layer 5: Upstream Systems (ERP + BOM)

Finally, go upstream:

- Was the work order correct?
- Was the BOM accurate?
- Was inventory properly defined?

➢ This is where the logic originates.

Why Layers Matter More Than Speed

New technicians try to:

- Fix fast
- Close tickets quickly
- Escalate when unsure

Experienced technicians:

- Slow down thinking
- Trace system logic
- Validate assumptions
- Confirm where the break actually started

Speed comes later.

Accuracy comes first.

⚠ The Common Trap: Jumping Too Early

Most mistakes happen when someone jumps:

- Straight to MES fix
- Or straight to system restart
- Or straight to escalation

Without asking:

"Which layer is actually responsible?"

Example of Wrong Thinking

"The lot is stuck, so MES must be broken."

That skips:

- Configuration
- Recipe linkage
- Tool eligibility
- ERP/BOM dependencies

Example of Layered Thinking

"The lot is stuck. Let me see what MES is expecting, and what rule is blocking movement."

That leads to:

- Faster root cause identification
- Fewer unnecessary escalations
- Better long-term understanding of the system

The Mental Model That Changes Everything

When something breaks, don't think:

"What is wrong?"

Think:

"At what layer is reality no longer matching expectation?"

That single shift changes everything about how you troubleshoot.

How Senior Technicians Actually Work

They don't chase errors. They trace logic:

- What did the system receive?
- What does it think should happen?
- What rule is enforcing that behavior?
- Where does that rule come from?

They are not reacting. They are *decoding*.

Simple Layer Summary

Layer	Focus	Question to Ask
1	Shop Floolr	What is physically happening?
2	MES Output	What does the system show?
3	MES Logic	What is MES expecting?
4	Configuration	How was MES set up to behave?
5	ERP/BOM	Where did the data originate?

The Real Skill of MES Work

It's not memorizing systems.

It's not clicking faster.

It's not knowing every screen.

It is:

The ability to see through layers of information until you find where alignment broke.

⚠ Why This Mindset Matters So Much

Without layered thinking:

- Everything feels like a system issue

- You rely heavily on escalation
- Problems take longer to solve
- You miss repeat patterns

With layered thinking:

- You solve issues faster
- You understand root causes
- You become the "go-to" person
- You start preventing problems, not just fixing them

Closing Line

MES problems are rarely where they appear.

The technician who understands the layers doesn't just fix issues—they understand them.

From Technician to Senior Technician: What Actually Changes in Your Thinking

Most people assume the difference is:

- More experience
- Faster troubleshooting
- Better system knowledge

That's partly true—but it's not the real shift.

The real change is **how you think about problems, not how many you've seen.**

1. From "Fixing Issues" → to "Understanding Behavior"

Technician mindset:

"Something is broken. Let me fix it."

You focus on:

- Errors
- Tickets
- Immediate resolution

Senior technician mindset:

"The system is behaving exactly as it was configured to behave. Why?"

Now you focus on:

- Rules
- Logic
- System behavior patterns

You stop treating MES like a broken system and start treating it like a **predictable system following instructions**.

2. From Single Problems → to System Patterns

Technician:

- Handles one issue at a time

- Sees each ticket as separate
- Focuses on the symptom in front of them

Senior technician:

- Notices repeating issues
- Recognizes patterns across tools, operations, or recipes
- Connects small issues to larger configuration gaps

Instead of:

"Why did this lot fail?"

They ask:

"Why are multiple lots failing in the same way?"

3. From Surface Thinking → to Layer Thinking (Automatically)

You already learned the layers earlier.

The difference is:

Technician:

- Uses layers when stuck

Senior technician:

- Uses layers by default

They automatically think:

- Shop Floolr → MES → configuration → ERP/BOM

Without being told.

It becomes instinct.

4. From "System Users" → to "System Interpreters"

Technician:

- Uses MES screens
- Follows procedures
- Executes tasks

Senior technician:

- Interprets what MES is *really saying*
- Understands why the system behaves a certain way
- Can explain issues in plain language to engineers, operators, and management

They don't just read errors.

They translate them.

5. From "Where is the issue?" → to "Where did the logic break?"

This is one of the biggest mindset shifts.

Technician:

"Where is the problem?"

Senior technician:

"Where did the logic or configuration diverge from reality?"

That means they think in terms of:

- Expectation vs reality
- Configuration vs execution
- Data vs physical behavior

Not just "what's wrong," but **why the system thinks it's right.**

6. From Reactive → to Preventative Thinking

Technician:

- Fixes what is broken right now
- Focuses on immediate production recovery

Senior technician:

- Asks what caused it
- Looks for upstream weaknesses
- Identifies repeat risk areas

They start preventing:

- Bad configurations
- Misaligned recipes
- Recurring tool issues

They don't just restore flow. They protect it.

7. From “System Knowledge” → to “System Ownership Thinking”

This is subtle but important.

Technician:

“I know how MES works.”

Senior technician:

“I understand how MES behaves when certain conditions exist.”

They stop memorizing and start **predicting**.

They can often anticipate:

- What will break next
- What change will cause downstream issues
- What configuration choice will create future bottlenecks

8. From Escalation Dependency → to Self-Reliance

Technician:

- Escalates when unsure
- Waits for engineering or IT confirmation
- Needs validation to proceed

Senior technician:

- Narrows down root cause before escalating

- Escalates with context, not confusion
- Often resolves issues before escalation is needed

They don’t just pass problems upward.

They **frame the problem correctly before it moves anywhere else.**

⚠ What Doesn’t Change (Important Reality Check)

Even at senior level:

- MES still breaks in confusing ways
- ERP issues still surface as MES problems
- Configuration mistakes still happen
- Shop Floolr pressure still exists

The difference is:

You don’t panic at the complexity anymore.

You recognize it as normal system behavior.

The Real Difference in One Line

Technician:

“Let me fix this issue.”

Senior Technician:

“Let me understand why the system is behaving this way—and whether it will happen again somewhere else.”

Final Insight

Becoming a senior MES technician is not about knowing more screens.

It's about seeing the system as a **connected logic chain**, not isolated problems.

Once that happens:

- Troubleshooting becomes faster
- Issues become predictable
- You stop reacting—and start interpreting

Chapter 9

MES Technician vs MES Engineer: What's Actually Different?

At first glance, the roles can look similar. Both:

- Work in MES systems
- Deal with production issues
- Interact with operations, recipes, and tools
- Support manufacturing execution

So it's easy to assume:

"They basically do the same thing, just at different levels."

But that's not accurate.

The difference is not just skill—it's **scope, ownership, and system control.**

MES Technician: Execution + Support Layer

Core Focus: Keeping production moving

An MES technician works closest to:

- The Shop Floolr
- Live production issues
- System behavior in real time

They typically:

- Monitor lots and operations
- Troubleshoot stuck or failed processes
- Validate configurations in practice
- Support operators and engineers
- Execute predefined procedures
- Identify where issues are occurring

How to Think About the Technician Role

A technician is primarily asking:

“What is happening right now, and how do I restore flow?”

They operate in:

- Real-time conditions
- Active production environments
- Immediate problem resolution

They are closest to:

Execution and reaction

MES Engineer: Design + System Ownership Layer

Core Focus: Designing how MES behaves

An MES engineer works more upstream in the system.

They typically:

- Design MES workflows and architecture
- Build or modify system logic
- Define integration with ERP and other systems
- Create or adjust configuration frameworks
- Develop standards for operations, recipes, and routing
- Troubleshoot complex system design issues
- Support long-term system stability and improvements

How to Think About the Engineer Role

An engineer is primarily asking:

"How should the system behave across all scenarios—and how do we design it correctly?"

They operate in:

- System design
- Logic creation
- Cross-functional planning
- Long-term stability and scalability

They are closest to:

Design and structure

Key Difference in One Line

- **MES Technician:** Fixes and supports what is happening now
- **MES Engineer:** Designs and defines how the system should behave

Where Confusion Happens in the Field

Here's where things get blurry in real environments:

Technicians often work with:

- Operations
- Recipes
- Tools
- Configuration checks

Engineers also work with:

- Operations
- Recipes
- Tools
- Configuration design

So from the outside, it looks identical. But the difference is: Technicians validate and apply the system. Engineers design and govern the system.

⚠ Why This Misunderstanding Causes Friction

Many technicians feel pressure because:

- They are expected to troubleshoot complex system logic
- They are asked to "fix configuration" without design authority
- They are pulled into issues that are actually architectural problems
- They are sometimes expected to behave like engineers without having control over system design

At the same time, engineers may assume:

- The system design is clear
- The configuration is correctly applied
- The execution layer will interpret logic correctly

When those assumptions don't align, gaps appear.

The Reality: They Are Connected, Not Equal

MES works because both roles exist:

Without technicians:

- Issues don't get resolved in real time
- Production stalls longer
- System behavior isn't validated on the floor

Without engineers:

- Systems become inconsistent
- Configuration becomes reactive instead of structured
- Long-term scalability breaks down

They are not the same role.

They are **two halves of system stability.**

A Simple Analogy

Think of MES like a building:

- **Engineer = Architect & Structural Designer**
  - Designs how the building should function
  - Decides structure, layout, systems
- **Technician = Facilities & Operations Support**
  - Ensures everything works day-to-day
  - Fixes issues as they appear
  - Keeps the building usable in real time

Both are essential—but they operate at different levels.

Final Insight

MES technicians and MES engineers work in the same system—but not at the same layer of control.

- Technicians live in execution and reality
- Engineers live in design and structure

The strongest MES environments happen when both understand each other's layer—not when one tries to perform both roles at once.

What "System Design" Means in MES

System design is deciding how MES should behave as a whole before people start using it.

It's the blueprint, not the day-to-day fixes—not the troubleshooting—but the structure that everything runs on.

Think of it like building a factory *inside the system*

System design answers questions like:

- How does a lot move from start to finish?
- What are all the required steps (operations)?
- What tools are allowed at each step?
- What rules must be satisfied before moving forward?
- How does MES talk to ERP and other systems?
- What data must be captured at each step?

It's the **architecture of the process inside MES.**

Simple definition

System design = how the MES process flow is structured and organized so production can run correctly.

What "Logic Creation" Means in MES

If system design is the blueprint, then:

Logic creation is writing the rules that control behavior inside that blueprint.

It defines *decision-making inside the system.*

Examples of MES logic

This is where you see rules like:

- "A lot cannot move unless recipe version X is active"
- "Only Tool Group A can run Operation 3"
- "If material is missing, place lot on hold"
- "If step is skipped, trigger validation error"
- "If ERP status is incomplete, block release to MES"

➢ These are not physical actions.

They are **system decisions based on conditions.**

Simple definition

Logic creation = defining the rules and conditions that determine how MES behaves in different situations.

How They Work Together

This is the key relationship:

System Design = Structure

- The flow of operations
- The setup of tools and recipes
- The overall process architecture

Logic Creation = Behavior

- The rules inside each step
- The conditions for movement
- The restrictions and validations

Easy analogy

Think of MES like a video game:

- **System design = the map**
- **Logic creation = the game rules**

Example:

- You can only enter a zone if you have a key → logic
- The map has 5 zones connected in a path → design

How This Shows Up in Real MES Work

When engineers do system design and logic creation, they are responsible for things like:

- Building routing flows for production
- Defining operation sequences
- Setting up tool restrictions
- Creating recipe-to-tool relationships
- Establishing validation rules for movement
- Connecting ERP data into MES behavior

⚠ Why This Is Often Confusing for Technicians

From a technician's perspective, you mostly see:

- A lot stuck
- A recipe error
- A tool mismatch
- A hold condition

But behind each of those is:

- A **logic rule being enforced**
- A **design decision made earlier**

So what looks like "a problem" is often:

a rule working exactly as it was designed to work

Key Insight

System design decides the path. Logic creation decides the rules on that path. MES execution follows both exactly.

What Strong MES Engineers Really Do

They don't just "build screens" or "set up tools."

They:

- Design how production flows should behave
- Anticipate where rules could conflict
- Create logic that prevents errors before they happen
- Ensure MES, ERP, BOM, and tools all align

Simple Summary

- **System Design = How MES is structured**
- **Logic Creation = How MES makes decisions**
- **Execution (Technician layer) = MES following those rules in real time**

⚠ Design vs Execution Mismatch: When the System Works "Correctly" but Production Doesn't

This is one of the most misunderstood realities in MES work.

A system can be:

- properly configured

- logically consistent
- technically "working as designed"

…and still cause real production problems. That's not a failure of execution. It's a mismatch between **design assumptions and real-world conditions.**

The Design (What Was Intended)

Engineering defines a process flow:

- A lot enters Operation A
- It uses Recipe Version 2.1
- It runs on Tool Group X
- It moves automatically to Operation B if all checks pass. On paper, everything is clean and efficient.

The logic is:

- controlled
- predictable
- repeatable

This is the **design layer**: how the system is supposed to behave

The Execution (What Actually Happens on the Floor)

Now the process runs in real production.

But reality introduces variables:

- Tool Group X is available, but partially qualified
- Recipe 2.1 exists, but operators are still using Version 2.0 habits
- A tool is down for maintenance, so alternate tools are used
- Inventory delays cause slight timing shifts in material availability

Now MES starts reacting to conditions the design didn't fully anticipate.

⚠ The Mismatch Appears

On the screen, everything looks like:

✓ Valid route

✓ Valid operation

✓ Valid recipe

✓ Valid tool group

But on the floor:

- Lots are slowing down
- Steps are intermittently blocking
- Operators are confused about eligibility
- Technicians are constantly clearing holds

So the question becomes:

"If everything is correct, why is everything stuck?"

Where the Real Problem Is

The issue is not a broken system.

It is a gap between:

Design assumptions:

- Ideal tool availability
- Perfect recipe adoption
- Stable production conditions
- Clean data flow from ERP

Execution reality:

- Tool variability
- Human behavior differences
- Timing delays
- Partial or inconsistent data alignment

Real Example of the Mismatch

Design says:

"Recipe 2.1 will always be used on Tool Group X."

Execution reality:

- Tool Group X is partially down

- Operators temporarily use Tool Group Y
- Recipe 2.1 is not fully validated on Tool Group Y

So MES does what it was told:

It blocks or restricts movement.

Not because it's broken—but because the execution no longer matches the original design assumptions.

What MES Is Actually Doing Here

MES is not deciding.

It is enforcing:

- design rules
- configuration logic
- validation conditions

So when execution deviates from design—even slightly—MES reacts consistently.

That consistency is often mistaken for "system issues."

The Technician's Real Challenge

From a technician's view, the frustration comes from this:

- The system looks correct
- The data looks correct
- But production behavior is not smooth

So the instinct is: "Something is wrong in MES."

But the deeper reality is:

"MES is still operating on assumptions that no longer match real conditions."

The Senior-Level Insight

Experienced MES technicians learn to ask:

"Is this a system failure—or a design assumption that no longer matches execution reality?"

That one question changes everything.

Because it shifts the focus from:

- fixing symptoms

to

- identifying structural mismatches

Why This Happens So Often

Design/execution mismatches usually come from:

- Over-idealized system design (too perfect assumptions)
- Incomplete Shop Floolr validation during setup
- Changes in tools or processes not fully reflected in MES
- Gradual drift between documented process and actual practice

Over time, small gaps accumulate. Until the system feels "off," even though nothing is technically broken.

Closing Insight (Book-End Message)

MES does not fail when it breaks. It reveals when design and execution no longer agree. The technician's job is not just to fix issues—but to recognize when reality has outgrown the original design.

Final Thought

The strongest MES professionals don't just ask:

- "What is wrong?"

They ask:

- "What assumption was this system built on—and is that still true today?"

That is the difference between:

- reacting to problems

and

- understanding systems

Chapter 10

Three Common MES Problems Every Technician Will Face

No matter how advanced the system is, real production environments are not perfect. As an MES technician, you will encounter issues that are not always caused by the system itself—but still require you to resolve them. Understanding these common problems early will help you respond faster and more confidently on the floor.

1. System Freezes and Slowdowns

What It Looks Like

- Screens stop responding
- Transactions take longer than expected
- Data does not update in real time
- Users repeatedly click or resubmit actions

What's Really Happening

A system freeze is not always an MES failure.

It is often caused by:

- Network latency or connectivity delays

- High system load (too many users or transactions at once)
- Integration lag between MES and other systems (like ERP)
- Background processing delays

Why This Matters

When the system slows down:

- Operators may assume nothing is happening
- Duplicate actions may be entered
- Data inconsistencies can occur
- Production flow becomes disrupted

What a Technician Should Do

- Verify if the issue is system-wide or isolated
- Avoid repeated inputs that could duplicate transactions
- Communicate clearly with operators ("the system is processing")
- Escalate if delays persist beyond normal thresholds

Key Insight

A slow system can create bigger problems than a stopped system if actions are repeated or entered incorrectly.

2. Barcode Scanner Issues

What It Looks Like

- Scanner won't read barcodes
- Incorrect data is scanned
- Delayed or inconsistent scans
- Manual entry becomes necessary

What's Really Happening

Barcode scanners are the bridge between the physical product and the digital system. When they fail, that connection is broken.

Common causes include:

- Damaged or poorly printed barcodes
- Scanner misalignment or hardware issues
- Connectivity problems
- User handling errors

Why This Matters

MES relies heavily on accurate scanning for:

- Lot tracking
- Material traceability
- Operation validation

When scanning fails:

- Data may be entered incorrectly
- Lots may be processed under the wrong conditions
- Traceability becomes unreliable

What a Technician Should Do

- Check the scanner hardware and connection
- Verify barcode readability
- Confirm correct data is being captured
- Limit manual entry when possible—and verify it carefully when required

Key Insight

If scanning is inaccurate, MES data becomes unreliable—no matter how well the system is configured.

3. Handling Rework and Scrap

What It Looks Like

- A lot fails a process step

- Material must be reprocessed
- Product is deemed unusable and must be scrapped
- System blocks forward movement

What's Really Happening

Rework and scrap are controlled events in MES—not informal decisions.

They are tied to:

- Quality rules
- Process validation
- System configuration
- Reporting requirements

MES enforces these actions to maintain:

- traceability
- accuracy
- accountability

Why This Matters

Improper handling of rework or scrap can:

- Distort production data
- Impact inventory accuracy
- Affect quality reporting

- Create confusion in downstream processes

What a Technician Should Do

- Follow defined MES procedures for rework or scrap
- Ensure the correct status is applied in the system
- Verify that routing or reprocessing steps are properly assigned
- Communicate with engineers or supervisors when rules are unclear

Key Insight

Rework and scrap are not just production decisions—they are system-controlled actions that affect the entire manufacturing record.

Closing Thought for This Section

The most common MES problems are not always system failures. They are breakdowns in interaction between people, tools, and data. A strong technician learns to recognize the difference—and respond accordingly.

MES Quick Reference:

Common Problems Checklist

Use this as a fast guide when issues happen on the Shop Floolr. Focus on identifying the problem quickly, stabilizing production, and avoiding additional errors.

System Freezes / Slowdowns

Check First

☐ Is the issue affecting one user or multiple users?
☐ Is the entire system slow, or just one screen/action?
☐ Has the action already been submitted (but not updated yet)?

⚠ **Avoid**

☐ Re-clicking or re-submitting transactions repeatedly
☐ Assuming the system is down without verifying
☐ Entering duplicate data

✓ **Do This Instead**

☐ Wait a few seconds and confirm system response
☐ Communicate with operators ("system is processing")
☐ Refresh or recheck status before taking action
☐ Escalate if delay continues beyond normal timing

Remember

Slow systems can cause duplicate or incorrect data if actions are repeated.

Barcode Scanner Issues

Check First

☐ Is the scanner powered on and connected?
☐ Is the barcode damaged or hard to read?
☐ Is the correct data being scanned into MES?

⚠ **Avoid**

☐ Blindly entering data manually without verification
☐ Assuming the scanner is working correctly without testing
☐ Skipping scans altogether

✓ **Do This Instead**

☐ Test the scanner on a known working barcode
☐ Clean or reprint damaged barcodes if needed
☐ Verify scanned data before submitting
☐ Use manual entry only when necessary—and double-check it

Remember

If the scan is wrong, the system data is wrong.

Rework & Scrap Handling

Check First

☐ Has the lot failed a defined process step?
☐ Does MES require rework routing or scrap confirmation?
☐ Are there specific rules tied to this operation?

⚠ **Avoid**

☐ Moving lots forward without proper system action
☐ Informally deciding rework or scrap without MES entry
☐ Skipping required validation steps

✓ **Do This Instead**

☐ Follow MES procedures for rework or scrap
☐ Apply the correct system status
☐ Confirm routing or reprocessing steps
☐ Notify engineer/supervisor if unclear

Remember

Rework and scrap must be recorded correctly—they affect traceability, inventory, and reporting.

Quick Troubleshooting Mindset

When in doubt, ask:

☐ What is happening on the floor?
☐ What is MES showing?
☐ What is MES expecting?
☐ Is this a system issue—or a data/configuration issue?

Bottom Line

Most MES problems are not system failures. They are interaction issues between people, tools, and data. Stay calm, verify before acting, and follow system logic.

MES QUICK REFERENCE SERIES

Use these as standalone pages or bundle them at the end of your book.

1. ERP Quick Reference (Upstream Control)

What ERP Does

☐ Creates work orders

☐ Defines production demand

☐ Manages inventory and materials

☐ Sends data to MES

How ERP Connects to MES

☐ ERP creates → MES executes

☐ Work orders flow into MES

☐ Inventory must match MES expectations

⚠ Common ERP-Related Issues

☐ Missing or incorrect work orders

☐ Inventory not available or misallocated

☐ BOM inconsistencies

☐ Timing delays between systems

✓ **What to Check**

☐ Is the work order released correctly?
☐ Is inventory available and accurate?
☐ Does the BOM match what MES expects?

Remember

If ERP data is wrong, MES will still execute it—incorrectly.

2. BOM Quick Reference (Material Structure)

What BOM Defines

☐ What materials are required
☐ Quantities needed for production
☐ Relationships between parts

How BOM Affects MES

☐ Controls material validation
☐ Drives inventory usage
☐ Determines what can move forward

⚠ Common BOM Issues

☐ Missing components
☐ Incorrect quantities

☐ Outdated structure

☐ Mismatch with MES configuration

What to Check

☐ Are all required materials present?

☐ Do quantities match production needs?

☐ Is BOM aligned with MES operations?

Remember

MES enforces the BOM—it doesn't question it.

3. Configuration Quick Reference (System Behavior)

What Configuration Controls

☐ Operations (process steps)

☐ Recipes (how steps run)

☐ Tool assignments

☐ Routing logic

☐ Validation rules

⚠ Common Configuration Issues

☐ Missing recipe-to-tool links

☐ Incorrect operation setup

☐ Incomplete routing

☐ Overly restrictive rules

What to Check

☐ Is the correct recipe assigned?

☐ Is the tool properly configured?

☐ Are all steps connected in the route?

Remember

MES is doing exactly what it was configured to do—even when it blocks production.

4. Troubleshooting Quick Reference (Layered Approach)

Always Follow This Order

☐ **1. Shop Floolr Reality**
What is physically happening?

☐ **2. MES Behavior**
What does the system show?

☐ **3. MES Expectation**
What is the system requiring?

☐ **4. Configuration Check**
How is it set up?

☐ 5. ERP/BOM Check

Where did the data come from?

⚠ Avoid

☐ Guessing

☐ Repeating actions

☐ Escalating without context

Remember

MES problems are usually misalignment—not failure.

5. Shop Floolr Issues Quick Reference

Common Issues

☐ System freezes

☐ Scanner failures

☐ Connectivity problems

☐ Data entry errors

Quick Actions

☐ Verify system vs local issue

☐ Check hardware (scanner, network)

☐ Confirm data accuracy

☐ Communicate clearly with operators

Remember

Many MES issues start outside the system.

6. Data & KPI Quick Reference

What MES Data Feeds

☐ Production output
☐ Downtime tracking
☐ Cycle times
☐ Scrap and rework

⚠ Common Data Issues

☐ Incorrect downtime codes
☐ Missing entries
☐ Delayed inputs
☐ Inconsistent logging

What to Do

☐ Enter data accurately and immediately
☐ Double-check critical inputs
☐ Follow standard categories

Remember

Your data becomes management decisions.

7. "Go-To Technician" Quick Reference

What Sets You Apart

☐ You think in layers
☐ You recognize patterns
☐ You stay calm under pressure
☐ You communicate clearly

Daily Habits

☐ Verify before acting
☐ Document issues clearly
☐ Learn from repeated problems
☐ Understand system behavior—not just screens

Remember

The best technicians don't just fix problems—they understand them.

Made in the USA
Monee, IL
08 November 2026

67378331R00090